3/11 25x

Black Widow Spiders

by Bill McAuliffe

Content Consultant:
Dr. Robert Breene
American Tarantula Society

RiverFront Books

An Imprint of Franklin Watts
A Division of Grolier Publishing
New York London Hong Kong Sydney
Danbury, Connecticut

RiverFront Books
http://publishing.grolier.com
Copyright © 1998 by Capstone Press. All rights reserved.
Published simultaneously in Canada.

Printed in the United States of America.

Library of Congress Cataloging-in-Publication Data
McAuliffe, Bill.
　　Black widow spiders/by Bill McAuliffe.
　　p. cm.--(Dangerous animals)
　　Includes bibliographical references and index.
　　Summary: Discusses the physical characteristics, habitat, behavior, and
lore of the infamous black widow spider.
　　ISBN 1-56065-619-0
　　1. Black widow spider--Juvenile literature. [1. Black widow spider.
2. Spiders.]　I. Title.　II. Series.
QL458.42.T54M39　　1998
595.4'4--dc21

　　　　　　　　　　　　　　　　　　　　97-8318
　　　　　　　　　　　　　　　　　　　　CIP
　　　　　　　　　　　　　　　　　　　　AC

Photo credits
Archive Photos/Lambert, 10, 14
James Cokendolpher, 8, 21, 22, 23, 28, 32
Leonard Lee Rue III, 12, 41
David Liebman, 6, 24, 42-43
Phillip Roullard, 18
John Serrao, 31, 39
Dan Suzio, 34
Visuals Unlimited/Adrian Wenner, 16; S. Maslowski, cover
Norbert Wu, 36

Table of Contents

Fast Facts about Black Widow Spiders

Scientific Grouping: Black widows belong to a genus of spiders known as widow spiders. A genus is a group of closely related plants or animals. There are 20 to 30 species of widow spiders. A species is a group of similar animals that can mate with each other.

Size: A black widow's legspan usually ranges from about one-half inch (one and one-half centimeters) to two inches (five centimeters). Males are usually smaller than females.

Coloring: Most black widows are a shiny black. Some species look purplish or brownish. Black widows have an hourglass-shaped marking on their bodies. Most markings are red, orange, or yellow. Sometimes the mark ranges in color from a dark reddish-brown to a very light tan.

Habitat: Black widows live on webs that they build in sheltered places. Black widows usually build their webs close to the ground.

Range: Five species of widow spiders live in North America. Only three of them are black widow species. These are the southern black widow, the northern black widow, and the western black widow. These spiders live in all but the most northern areas of North America. Some live in South America and Central America, too.

Food: A black widow mostly dines on a variety of insects. But it can also kill and eat larger prey such as lizards, frogs, mice, and snakes.

Behavior: Black widows are predators. A predator is an animal that hunts another animal for food.

Chapter One

The Black Widow Spider

Black widows are predators. A predator is an animal that hunts another animal for food. Black widows repeatedly bite their victims with their fangs. A fang is a long, pointed tooth with a small hole in its end.

But it is not just black widow fangs that kill victims. Black widows produce a poison called venom. Some black widows have venom that is 15 times more poisonous than rattlesnake venom. Black widow venom contains poisons that attack nerves. When black widows bite, they release venom into their victims' bodies. Most of the time, the venom kills the victims.

Black widows are predators.

Black widows' venom attacks the nerves of its victims.

In the past, one out of every 25 people bitten by black widows died. Those who died were usually small children, older people, or people already in poor health.

Black Widow Bites
Black widow bites produce pain at the places where people are bitten. This lasts for several hours. Then the victims become sick to their stomachs. They become dizzy and sweat a lot.

As the venom attacks the nerves, victims' talking and breathing may be affected. Within three days, jaw and stomach muscles might go into painful spasms. A spasm is an uncontrollable muscle tightening.

In the past, doctors did not recognize the signs of black widow bites. They thought the victims had other diseases. Sometimes the doctors performed surgeries. The surgeries sometimes killed the victims.

Today, an antitoxin is available that quickly eases the problems caused by a bite. An antitoxin is a special medicine that fights poisons. Most victims recover fully from black widow spider bites. People should go to the hospital if they believe they have been bitten by a black widow.

It is true that black widow bites can be dangerous. But people's chances of being bitten by black widows are less than the chances of being struck by lightning. This is because black widows bite people only if the spiders are dying or hurt. They may also bite if

Female black widows are the largest spiders in the cobweb weaver family.

they feel danger or mistake people for prey. Prey is an animal hunted or eaten as food.

Black widows are a much more serious threat to neighboring insects. They also kill mice and small snakes that stumble into their webs.

So far, only females have been reported as biting people. Males have the same venom, but they have never been seen biting people.

Range

Black widows belong to a genus of spiders known as widow spiders. A genus is a group of closely related plants or animals. There are 20 to 30 species of widow spiders. A species is a group of similar animals that can mate with each other.

Five species of widow spiders live in North America. Only three of them are black widow species. These are the southern black widow, the northern black widow, and the western black widow. These spiders live in all but the most northern areas of North America. Some live in South America and Central America, too.

Appearance

Widow spiders are cobweb weavers. Cobweb weavers make webs. Female black widows are the largest spiders in the cobweb weaver family. Their legspan usually ranges from about one-half

inch (one and one-half centimeters) to two inches (five centimeters). Males are usually smaller than females.

Most black widows are a shiny black. Some species look purplish or brownish. Spiderlings have lighter colorings such as orange or white. A spiderling is a young spider. Spiderlings' colors grow darker as they age.

People can easily recognize black widows. They have an hourglass-shaped marking on the underside of their bodies. Sometimes the hourglass appears as two triangles. Most markings are red, orange, or yellow. Sometimes the mark ranges in color from a dark reddish-brown to a very light tan.

Male black widows are brown or cream colored. They have a pair of reddish lines along each side.

Most parts of a black widow's body are covered with an exoskeleton. An exoskeleton is a structure on the outside of an animal. Some

A black widow has an hourglass-shaped marking on the underside of its body.

parts of the exoskeleton are hard and some parts are soft. It helps a spider keep water in its body. An exoskeleton also helps protect a black widow from its enemies.

The Cephalothorax

A black widow shares some features with all spiders. Its body has two main parts. The front part is the cephalothorax. The cephalothorax contains many organs. These include a black widow's brain, eyes, head, fangs, and mouth.

One chelicera is above each side of a black widow's mouth. A chelicera is a body part by a spider's mouth that is used like a jaw. Fangs are a part of the chelicerae.

A special arm-like head segment called a pedipalp is next to each chelicera. The pedipalps are much smaller than a black widow's legs. A black widow sometimes uses pedipalps to position its prey for eating. Special hairs that help a black widow smell and taste are on the end of each pedipalp.

A black widow's body has two main parts.

All spiders have eight legs that attach to the cephalothorax. People have joints at the hip, knee, and ankle. Each of a black widow's legs has seven joints. It can bend its legs at each of the joints.

The Abdomen

The second part of a black widow's body is the abdomen. It is in back of the cephalothorax. The abdomen is shaped like a bulb. It contains many organs including the hearts, lungs, and spinnerets.

A spinneret is a body part at the rear of the abdomen. A black widow's body produces liquid silk. The spider uses its spinnerets to spin this silk into threads and webs. The silk is very strong. It traps and holds captured prey.

Strong Silk

Black widows can make two kinds of silk. One kind is four times stronger than the silk of any other spider.

In fact, black widow silk is as strong as Kevlar. Kevlar is a bullet-proof material. Some scientists think people could use black widow silk to make bullet-proof vests.

Each of a black widow's legs has seven joints in it.

Chapter Two

Survival

Black widows are cold-blooded. Cold-blooded means that black widows' body heat comes from their surroundings. Warm areas are ideal for black widows to make their homes.

Webs

A black widow lives in a web. It makes the web with its spinnerets.

A black widow web is not as fancy as other spiders' webs. Instead, the web is a tangle of strong silk. Tensile strength is the amount of weight something can bear without tearing apart. The silk has the tensile strength of steel.

A black widow's web can be two to three feet (60 to 90 centimeters) wide. Sometimes

Black widows live in webs they make with their spinnerets.

the web can be larger. The size of the web often depends on the amount of space available to the spider.

A black widow finds a sheltered place to build its web. This place is usually close to the ground. A web location must have a narrow, funnel-shaped opening. This opening can be natural or made by people. A black widow uses this opening as a retreat. A black widow builds its web around this opening and hides inside it.

People find black widow webs under boards or rocks. Some webs are in burrows or tree hollows. A burrow is a hole dug by an animal to use as a home.

Black widows hang upside down from their webs. They leave their webs if an enemy moves nearby. They also move if they are unable to catch any food. Males leave their webs to find females to mate with.

Catching Prey

A black widow mostly dines on a variety of insects. But it can also kill and eat larger prey such as lizards, frogs, mice, and snakes.

Black widows hang upside down from their webs.

Black widows trap prey in the sticky silk of their webs.

Sometimes a black widow scavenges for food. A scavenger is an animal that searches and eats another animal's leftover food.

Most of the time, a black widow uses the sticky silk of its web to catch prey. It hides itself in its retreat. It waits for prey to become trapped in the web. A black widow has bad eyesight. So its web serves as an alarm that lets it know when prey is present. When the prey

Black widows bite their prey and wrap prey in silk.

touches the web, the web moves. The black widow senses the movement. Then the black widow climbs onto its web.

The black widow bites the prey with its fangs. It begins to spin silk and wrap it around the prey. As the prey struggles, the black widow continues to bite until the prey no longer fights.

Eating

Many black widows stay on their webs to eat. But some carry the prey back to their retreats. Black widows can be very strong. One black widow was seen lifting a small lizard. It carried the lizard more than three feet (one meter) to its retreat. A black widow in the United States was photographed with a small snake. It had wrapped silk around the snake. It bent the snake's head to its tail. Then the black widow carried the snake back to its retreat.

Like all spiders, black widows can only swallow liquids. They digest their food outside of their bodies. Their mouths release special digestive juices that dissolve prey. Then black widows drink the liquids from the prey. Black widows have special organs around their mouths. These organs filter out larger bits of prey.

Enemies

Wasps, praying mantises, and some other spiders are the enemies of black widows. Mud dauber wasps also attack black widows.

Praying mantises are enemies of black widows.

A mud dauber wasp uses black widows to help it breed. It first stings the black widow. The mud dauber wasp injects its venom into the black widow. The venom makes the black widow unable to move. The wasp then carries the black widow to a mud cell.

The wasp lays an egg on the first black widow it puts in the cell. It then catches from three to nine other spiders and puts them in the cell, too. All the black widows are still alive, but unable to move. The mud dauber wasp seals the living spiders in the cell with its egg.

When the egg hatches, the mud dauber wasp larva eats the black widows alive. A larva is a young insect during its worm-like growth stage. The black widows die when the larva eats an important organ. Meanwhile, the black widows are unable to move or defend themselves. Black widows might stay alive for weeks while they are being eaten. All the spiders in the mud cell are eaten by the growing wasp.

Mud dauber wasps put black widows in mud cells.

Chapter Three

A Black Widow's Life

Humans have an inside skeleton covered with soft skin. The skin stretches and allows growth. Black widows' hard exoskeletons do not stretch and grow. Growing black widows must shed their exoskeletons. This process is called molting.

Black widows' old exoskeletons peel from their new ones during the molt. This can take several hours. Black widows are very weak during a molt. They do not have their hard covering to protect them. It takes a while for the new exoskeleton to harden. The new exoskeleton is larger than the old one. This lets black widows grow.

Black widows shed their exoskeletons during a molt.

Newborn spiderlings molt every other week. A spiderling is a young spider. Then they molt every month until they are fully grown. After they become an adult, they never molt again.

During a molt, black widows can grow new legs. They do this if they have missing or seriously damaged legs.

After reaching adulthood, males live about 127 days. Females can live up to 849 days after becoming an adult. Black widows do not mate until they reach adulthood.

Mating

Male and female black widows usually live alone. This makes mating an unusual event in many ways.

Females' bodies produce special chemicals when they are ready to mate. These chemicals are called pheromones. Scientists believe that females put pheromones on their webs. Pheromones release scents that may attract males and help them find the females.

Mating can be an unusual event.

A male calms a female by tapping the female's legs.

A male senses many things from the pheromones in a female's web. The pheromones tell the male which behaviors he needs to perform before he can mate.

First, the male begins vibrating his abdomen. The vibrations send signals to the female. The signals calm the female and lessen the chance that she will attack the male.

As the male vibrates, he begins cutting the female's web. He cuts away parts of the web. Then he spins silk and wraps the parts of the female's web into bands and balls. Poor eyesight makes the female depend upon touching her web to know her location. When a male cuts the web, scientists believe it may be like blinding the female.

The male approaches the female when he is finished cutting the web. The male spins thin threads of silk over her. The threads form what is called a bridal veil. The veil is not like a trap. A female could easily break the threads in the bridal veil. Scientists believe the veil might be used to further calm the female.

After the veil is spun, the male usually begins tapping the female's legs and then her abdomen. This further calms the female. Then the black widows mate.

After mating, the male may quickly leave the female's web. He might stay near the female's web, or he might search for another female to mate with. Most people believe that the female eats the male after mating. But only two widow species do this.

Black widows lay about 250 eggs in an eggsac.

Sometimes a female may refuse all of the male's attempts to mate. The male might stay near her web for several weeks and continue trying to mate. Other times he might leave.

Eggs

Female black widows commonly lay eggs about every month. They spin silk to make an eggsac. This protects the eggs. Black widows can lay about 250 eggs in an eggsac. The eggs hatch in about 20 days. Not all of these eggs hatch. Some are eaten by insects.

The newly hatched spiderlings do not come out of the eggsac immediately. The spiderlings grow in the eggsac for several weeks. They molt two or three times while they are in the eggsac. They eat the egg yolk while they are inside the eggsac.

After the spiderlings have grown, they leave the eggsac. They spin silk threads that become caught in the wind. The threads lift spiderlings so that they float into the air. The spiderlings spin webs and live wherever they land. This process is called ballooning.

Sometimes the spiderlings are unable to open the eggsac. When they are trapped, they might eat their siblings for food. A sibling is a brother or sister.

Chapter Four

Black Widows and People

John Henry Comstock was one of the first entomologists to talk and write about black widows. An entomologist is a scientist who studies insects.

Comstock published *The Spider Book* in 1912. In this book, he called the genus of spiders black widows. Comstock was probably the first person to use the term black widow.

Name

Comstock named the spiders black widows because he believed females ate males after

Comstock believed that female black widows ate the males.

mating. Comstock saw this behavior because he put a male and female widow in an escape-proof cage. Since the spiders could be dangerous to people, he did not want them to escape.

The black widows mated. After mating, the male could not leave. After a time, the female found the male and ate him. The female probably ate the male because of the escape-proof cage.

In the wild, females do not usually eat the males. Only females of two widow species eat the males. Females of other widow species only eat the males if the males are weak or sick. So most males are not in danger after mating.

Entomologists and the Black Widow

For many years, entomologists put black widows in escape-proof cages to study their mating behavior. They believed that males could only mate once. They also believed that most females ate the males.

In the 1960s, entomologists changed the genus name black widow. The Committee on

Some species of widow spiders are not black.

Common Names of Arachnids of the American Arachnological Society made the decision. Members of this society study arachnids. Arachnids are a group of animals that includes all spiders, as well as scorpions, mites, and ticks. They dropped the name black from the genus name. Today, the genus is called widow spiders. They changed the name because some species of widow spiders are not black.

In the 1980s, scientists discovered that males could mate more than one time. They discovered that females rarely eat males. But the name widow was not changed because most people know the name.

Future

People need spiders to help keep balance in nature. Black widows eat many insects that hurt plants. They also serve as food for other animals like birds and lizards.

Learning about black widows will help people understand the value of these spiders. If people treat the spiders carefully, they can avoid being bitten. Black widows do not plan attacks on people. If people learn to respect black widows, the spiders will continue to play an important role in the natural world.

Spiders help keep balance in nature.

Hourglass-shaped marking

Abdomen

Spinnerets

Jointed Leg

Cephalothorax

Pedipalp

Chelicera with
fang

Words to Know

abdomen (AB-duh-muhn)—the rear part of a spider's body

antitoxin (an-ti-TOK-sin)—a special medicine that fights poisons

ballooning (buh-LOON-ing)—a method many spiderlings use for traveling; they are carried in the wind while suspended from a silk thread.

cephalothorax (se-phuh-luh-THOR-aks)—the head and middle part of a black widow's body

chelicera (ki-LI-suh-ruh)—a body part near a spider's mouth that is used like a jaw

exoskeleton (eks-oh-SKEL-uh-tuhn)—a hard structure on the outside of an animal

fang (FANG)—a long, pointed, hollow tooth

molt (MOHLT)—the process of shedding an old exoskeleton

spiderling (SPYE-dur-ling)—a young spider

venom (VEN-uhm)—a poisonous liquid produced by some animals

To Learn More

Gerholdt, James E. *Black Widow Spiders*. Edina, Minn.: Abdo & Daughters, 1996.

Martin, Louise. *Black Widow Spiders*. Vero Beach, Fla.: Rourke Enterprises, 1988.

Murray, Peter. *Black Widows*. Mankato, Minn.: Child's World, 1992.

Nielsen, Nancy J. *The Black Widow Spider*. New York: Crestwood House, 1990.

Useful Addresses

American Arachnological Society
Department of Entomology
The American Museum of Natural History
Central Park West at 79th Street
New York, NY 10024

Central California Arachnid Society
4082 North Benedict #104
Fresno, CA 93722-4559

Smithsonian Institution
Office of Elementary & Secondary Education
Smithsonian Institution
Washington, DC 20560

Young Entomologists' Society, Inc.
1915 Peggy Place
Lansing, MI 48910-2553

Internet Sites

Arachnids: Black Widow Spider
http://caboose.com/a1topics/portland_zoo/
   maingate/insect_zoo/Arachnids_Black_Widow

Arachnology Page
www.ufsia.ac.be//Arachnology/Arachnology.html

Black.html
http://www2.d25.k12.id.us/~rudeer/black.html

The Bug Club Home Page
http://www.ex.ac.uk/bugclub/welcome.html

Index